Henry C. Burdett

The Relative Mortality After Amputations of Large and Small Hospitals

And the influence of the antiseptic (Listerian) system upon such mortality

Henry C. Burdett

The Relative Mortality After Amputations of Large and Small Hospitals
And the influence of the antiseptic (Listerian) system upon such mortality

ISBN/EAN: 9783337239756

Printed in Europe, USA, Canada, Australia, Japan

Cover: Foto ©berggeist007 / pixelio.de

More available books at **www.hansebooks.com**

THE

RELATIVE MORTALITY

AFTER

AMPUTATIONS

OF

LARGE AND SMALL HOSPITALS,

AND THE INFLUENCE OF

THE ANTISEPTIC (LISTERIAN) SYSTEM

UPON SUCH

MORTALITY.

By HENRY C. BURDETT,

FELLOW OF THE STATISTICAL SOCIETY; HONORARY SECRETARY, HOME HOSPITALS'
ASSOCIATION FOR PAYING PATIENTS; LATE SECRETARY AND GENERAL
SUPERINTENDENT OF THE SEAMEN'S HOSPITAL, GREENWICH; THE
QUEEN'S HOSPITAL, BIRMINGHAM, ETC., ETC.

REPRINTED FROM THE JOURNAL OF THE STATISTICAL SOCIETY, SEPTEMBER, 1882.

LONDON:
J. AND A. CHURCHILL, 11, NEW BURLINGTON STREET.

1882.

. *Long experience of hospital management has led me to the conclusion that the laymen, who we must remember constitute by far the larger number of those who are engaged in hospital administration, realise far too little the importance of the hygienic surroundings, and the general habit—if I may use that word—of hospitals and hospital wards. I therefore believe and hope that by the publication of this review of an interesting question in relation to hospital management, I shall not only attract to it the attention of the medical profession, which has already indirectly been drawn to this subject by the medical press, but that I shall reach a wider audience, and possibly some at least of those earnest laymen who devote so much time and attention to the wise administration of our hospitals. If I succeed in this, I may in some way aid the medical profession by directing the attention of such laymen to the important questions with which I have dealt in the following pages. I know as a fact, from the investigations I have made, that comparatively few hospitals, large or small, possess an accurate and reliable plan of their drainage arrangements,—a fact to which I desire once more to direct the attention of the Medical Staff and Committees of Management.*

H. C. B.

THE RELATIVE MORTALITY

AMPUTATIONS

OF

LARGE AND SMALL HOSPITALS.*

PROBABLY no question has been more keenly contested than the one which constitutes the heading of this paper. Ever since Sir James Simpson published his famous essay on the subject the controversy has continued, and the disputants on either side have held to their opinions often with considerable warmth. In fact, the discussion of the relative mortality of large and small hospitals has generated more heat at times than can be easily accounted for.

The accuracy of Sir James Simpson's statistics of the results of amputations in country and private practice has been seriously impugned by Callender, Holmes, and other authorities, owing to the impossibility of proving the reliability of the sources from which they were derived, and because no details of the cases were given.

Feeling deeply the importance of the subject, it seemed to me of interest to collect actual figures, which could be definitely verified from the books kept by the medical staff of the different hospitals. With this view a circular, a copy of which is given in Appendix A, p. 35, was despatched to 160 cottage hospitals.

Answers were received in reply from ninety-two cottage hospitals, into thirty-one of which no cases requiring amputations had been received, although the majority had had, in addition to severe fractures, cases of herniotomy, lithotomy, extirpation of eyeball, removal of bone for necrosis, ovariotomy, excisions of knee, ankle, shoulder and breast, or other surgical cases of interest.

* The operations tabulated in this Paper comprise all that are recorded in the Cottage Hospital Case Books from the date of their foundation to the end of the year 1878, a period of twenty years.

Table of Amputations and their Results, Primary or for Injury, and Secondary or for Disease, of the Thigh, Leg, Arm, and Forearm, performed in Cottage Hospital Practice by Country and Provincial Practitioners.

Nominal List of Cottage Hospitals	Number of Beds	Primary								Secondary								Total	
		Thighs		Legs		Arms		Forearms		Thighs		Legs		Arms		Forearms			
		C.*	D†	C.	D.	C.	D.	C.	D.	C.	D.	C.	D.	C.	D.	C.	D	C.	D.
Ashford	6	1	1	5	1	–	–	2	–	2	–	–	–	–	–	–	–	10	2
Beccles	7	–	–	1	–	–	–	–	–	–	–	1	–	–	–	–	–	1	–
Bournemouth	6	–	–	1	–	2	–	–	–	–	–	–	–	–	–	–	–	3	–
Burton-on-the-Water	8	1	1	–	–	2	–	1	–	–	–	–	–	1	–	1	–	5	1
Burford	6	–	–	1	–	1	–	–	–	–	–	–	–	–	–	–	–	2	–
Bromley	10	–	–	–	–	1	–	–	–	1	1	–	–	–	–	1	–	3	1
Boston	5	1	–	1	–	2	1	1	1	–	–	–	–	–	–	–	–	5	2
Buckhurst Hill	7	–	–	–	–	1	1	–	–	–	–	–	–	–	–	–	–	1	1
Bromyard	5	–	–	–	–	–	–	1	–	–	–	2	–	–	–	–	–	3	–
Crewkerne	12	1	1	1	1	3	–	2	–	–	–	–	–	–	–	–	–	7	2
Cromer	6	–	–	–	–	–	–	–	–	1	–	–	–	–	–	–	–	1	–
Chesham	7	–	–	–	–	–	–	–	–	1	–	–	–	–	–	–	–	1	–
Cranleigh	6	1	–	–	–	–	–	1	1	3	–	–	–	2	–	–	–	7	1
Cirencester	6	–	–	–	–	–	–	1	–	4	–	–	–	1	–	–	–	6	–
Charlwood	4	–	–	1	1	–	–	–	–	–	–	–	–	–	–	–	–	1	1
Dorking	12	–	–	–	–	1	–	–	–	1	–	–	–	–	–	–	–	2	–
Enfield	6	–	–	–	–	–	–	–	–	1	–	–	–	–	–	–	–	1	–
Erith	7	–	–	–	–	–	–	–	–	1	–	2	–	–	–	–	–	3	–
Fairford	8	–	–	–	–	–	–	1	–	–	–	–	–	–	–	–	–	1	–
Fowey	8	1	–	1	–	–	–	–	–	–	–	3	–	–	–	–	–	5	–
Frome	10	–	–	–	–	1	–	–	–	–	–	–	–	–	–	–	–	1	–
Hayes	5	–	–	–	–	–	–	1	–	–	–	–	–	–	–	–	–	1	–
Hillingdon	4	–	–	–	–	1	–	–	–	–	–	–	–	–	–	–	–	1	–
Hatfield Broad Oak	8	–	–	–	–	–	–	1	–	1	1	–	–	–	–	–	–	2	1
Jarrow-on-Tyne	11	3	2	6	1	2	–	–	–	–	–	–	–	–	–	–	–	11	3
Iver	7	–	–	–	–	–	–	1	–	–	–	–	–	–	–	1	–	2	–
Kendal	16	–	–	–	–	1	1	1	–	3	1	3	–	1	–	–	–	9	2
Ledbury	5	–	–	1	–	2	–	–	–	1	–	–	–	–	–	1	–	5	–
Lloyd (Bridlington)	13	–	–	2	–	–	–	–	–	3	1	–	–	–	–	–	–	5	1
Litcham	7	1	1	–	–	–	–	–	–	–	–	–	–	–	–	–	–	1	1
Marlborough (Savernake)	20	3	–	4	–	2	–	1	–	1	–	–	–	1	–	1	–	13	–
Mildenhall	8	3	2	2	–	1	–	–	–	2	–	–	–	1	–	–	–	9	2
Malvern	12	–	–	2	–	–	–	–	–	5	–	3	–	–	–	–	–	10	–
Market Rasen	4	–	–	–	–	–	–	–	–	–	–	1	–	–	–	–	–	1	–
Milton Abbas	6	–	–	–	–	–	–	–	–	–	–	–	–	1	–	–	–	1	–
Melksham	6	1	1	–	–	–	–	1	–	–	–	–	–	–	–	–	–	2	1
Newton Abbot	8	–	–	–	–	2	1	–	–	–	–	–	–	–	–	–	–	2	1
North Cambridgeshire	23	–	–	–	–	1	–	–	–	2	1	–	–	–	–	–	–	3	1
North Lonsdale	50	5	4	10	–	8	1	–	–	–	–	1	–	–	–	–	–	24	5
Ottery St. Mary	7	–	–	1	–	–	–	3	–	2	–	1	1	1	–	–	–	8	1
Oswestry	8	2	–	2	–	3	1	2	–	–	–	–	–	–	–	–	–	9	1
Petworth	8	–	–	–	–	–	–	–	–	1	1	–	–	–	–	–	–	1	1
Petersfield	6	1	–	–	–	–	–	–	–	–	–	–	–	–	–	–	–	1	–
Penrhyn	13	–	–	–	–	1	–	–	–	–	–	–	–	–	–	–	–	1	–
Ross	6	–	–	–	–	–	–	2	–	–	–	–	–	–	–	–	–	2	–
Royston	7	1	1	–	–	1	–	–	–	3	–	3	1	–	–	3	–	11	2
Reigate and Redhill	12	–	–	1	1	2	–	1	–	1	1	2	–	–	–	–	–	7	2
Rugeley	10	1	1	2	1	–	–	1	–	2	1	–	–	–	–	–	–	6	3
Ross Memorial (Dingwall)	4	–	–	–	–	–	–	1	–	–	–	–	–	1	–	–	–	1	–
Ruabon	6	1	1	4	–	–	–	–	–	–	–	–	–	1	–	–	–	6	1
Stockton-on-Tees	9	1	–	23	7	7	1	5	–	2	1	2	1	–	–	–	–	40	10
St. Albans	7	1	1	2	–	–	–	3	–	1	–	–	–	–	–	–	–	7	1
Seacombe	8	1	1	1	1	–	–	–	–	–	–	–	–	2	–	–	–	4	2
Stratton (Cornwall)	5	–	–	–	–	1	–	–	–	–	–	–	–	–	–	–	–	1	–
St. Leonards (Sudbury)	20	2	1	–	–	2	–	2	–	2	–	1	–	–	–	1	–	10	1
St. Mary's (Burford Tenbury)	8	–	–	–	–	–	–	1	–	–	–	–	–	–	–	–	–	1	–
Tewkesbury	9	1	1	5	–	4	–	5	–	6	–	5	–	–	–	–	–	26	1
Trowbridge	8	–	–	–	–	–	–	3	1	–	–	1	–	–	–	–	–	4	1
Ulverston	12	2	1	2	1	–	–	2	–	1	–	–	–	1	–	–	–	7	2
Warminster	7	–	–	–	–	–	–	–	–	–	–	1	–	–	–	–	–	1	–
Wickham	6	–	–	1	–	–	–	–	–	–	–	–	–	–	–	–	–	1	–
	551	36	21	82	15	55	7	47	3	54	9	32	3	11	–	9	–	326	58

* C. cases.   † D. deaths.

The cases of amputation in the sixty-one hospitals, which are given in alphabetical order in the following table, amount to 326, or nineteen more than the number given by Professor Erichsen in his book, as "all the amputations which have been performed in "his wards at University College Hospital from the foundation of "the hospital, a period of thirty-eight years." The average mortality in Professor Erichsen's cases was 25 per cent., while it amounted to a little over 17 at these cottage hospitals.

To facilitate comparison, the following summary of the above table has been prepared on the plan adopted by Sir James Simpson :—

1. *Total mortality of all amputations in sixty-one cottage hospitals, having a total of 553 beds.*

Total number of cases, 326.

„   deaths, 58.

Or 1 in every 5·6 died ; or 17 in every 100.

2. *Mortality of the individual amputations.*

When we include all the amputations of the thigh, leg, arm, and forearm, the results are :—

Thigh   cases   90 ; deaths 30 ; or 1 in   3   ; or 33·3 per cent.
Leg   „   114 ;   „   18 ;   „   6·3 ; „ 15·5   „
Arm   „   66 ;   „   7 ;   „   9·4 ; „ 10·6   „
Forearm „   56 ;   „   3 ;   „   18·6 ; „ 5·4   „

3. *Mortality from the amputations that were primary or for injury.*

Thigh   cases 36 ; deaths 21 ; or 1 in   1·7 ; or 58 per cent.
Leg   „   82 ;   „   15 ;   „   5·5 ; „ 18   „
Arm   „   55 ;   „   7 ;   „   6·6 ; „ 15   „
Forearm „   47 ;   „   3 ;   „   15·6 ; „ 6·4   „

4. *Mortality from the amputations that were secondary or for disease.*

Thigh   cases 54 ; deaths 9 ; or 1 in 6   ; or 16·6 per cent.
Leg   „   32 ;   „   3 ;   „ 10·7 ; „ 9   „
Arm   „   11 ;   „   *nil.*
Forearm „   9 ;   „   *nil.*

These tables will be incomplete unless the cause of death in each case is recorded. Thus, in the primary amputations for injury :—

Of the thigh cases—13 died from shock.
1   „   pyæmia.
1   „   enteritis.
1   „   inflammation of lungs.
1   „   delirium tremens.

B

In the remaining case—a compound fracture just above the knee, with destruction of the femoral artery, not detected at the time of reduction,—mortification of the limb set in, and amputation was performed as the last resource.

Of the leg cases—6 died from shock.

3	,,	pyæmia.
1	,,	tetanus.
1	,,	delirium tremens.
1	,,	pneumonia.
3	,,	not stated.

Of the arm cases—4 died from shock.

1	,,	pneumonia.
1	,,	tetanus.
1	,,	not stated.

Of the forearm cases—2 died from shock.

1	,,	tetanus.

In the secondary amputations for disease :—

Of the thigh cases—3 died from exhaustion.

2	,,	secondary hæmorrhage.
1	,,	shock.
1	,,	pyæmia.
2	,,	not stated.

Of the leg cases—2 died from exhaustion.

1	,,	not stated.

The cases in which the cause of death is not stated were treated at the Stockton Hospital, the books of which give no information on the point. Of the five cases of pyæmia, two occurred at Stockton, one at Crewkerne, one at Ashford, and one at the Lloyd Cottage Hospitals.

It will be observed that the great mortality in the primary amputation of the thigh is due to the fact that four-fifths (17) of the deaths were caused by shock, consequent upon the severe injuries which the patients had sustained.

I have shown in the above table that the mortality after amputations in cottage hospital practice, in hospitals having 553 beds, is 17 per cent. In four leading metropolitan hospitals, containing upwards of 1,800 beds, Professor Erichsen[*] shows the mortality, after operations, to have been 37·8 per cent. The mortality in the Parisian hospitals,[†] as given by Malgaigne and Husson, Holmes and Bristowe, amounts to 60 per cent. Billroth[‡] gives the mortality at Zurich, between the years 1860 and 1867, as 46 per cent. Sir James Simpson[§] gives the mortality in town hospitals after these cases as 41·6 per cent.; at the Edinburgh

[*] "On Hospitalism," p. 20. Longmans, 1874.
[†] Ibid., p. 11.
[‡] Billroth, "Chirurgische." Klinik, Zurich, 1866-67.
[§] Simpson's Works, vol. ii, pp. 280—400; Article "Hospitalism."

Infirmary as 43·3 per cent.; at the Glasgow Infirmary as 39·1 per cent.; at St. Bartholomew's Hospital as 36·6 per cent.; at the London Hospital, Whitechapel, as 47·3 per cent.; at Guy's Hospital as 38·2 per cent.; and at St. George's Hospital 38·8 per cent. These statistics were brought down to the year 1868, and, as Professor Erichsen* truly says, "the accuracy of Sir James Simpson's " statistics relating to hospital practice has been admitted by all, " even by his most determined opponents, for they have been " derived from statistical returns furnished to him by the surgeons " and registrars of the various hospitals to which they relate." In the result the lowest mortality in any of the metropolitan hospitals referred to by Sir James Simpson is 34·4 per cent., the highest 47·3 per cent.; whereas the cottage hospitals show an average mortality of but 17 per cent.

When I began to collect materials for this paper, in 1876-77, I thought it would suffice to publish the foregoing tables and remarks. I was soon, however, undeceived. Certain critics threw doubts upon the value of my tables and statistics, on the ground that they were figures and figures only.

It was argued that the question at issue was mis-stated by me. Objection was taken to my figures, as to Simpson's, because they are unaccompanied by "any facts, any particulars of the " cases, and are therefore susceptible of any number of different " interpretations besides the one which Simpson chose to select, " viz., that there is an inherent unhealthiness in large hospitals, " which he described by the term ' hospitalism.' " It was alleged that the difference in favour of cottage hospitals of 7 per cent. in the number of deaths after amputations of the limbs, "may as " easily have depended upon difference in the surgical practice, in " the vitality (from age, state of health, &c.) of the patients, or on " the previous condition of disease or injury, or in fact on any " conceivable combination of these, and very possibly of other " causes, as on a difference in the healthiness of the hospitals." One critic, in fact, congratulated the large hospitals on the fact that a difference in mortality of 7 per cent. "proves that the intrinsic " danger of operations in cottage or in large hospitals cannot be " great." As to this, it is only to be observed that a death-rate of 70 per 1,000 in any community would hardly be regarded as a trifle, even by the most indifferent of sanitarians. After regretting " the absence of any attempt to estimate the real sanitary condition " of cottage hospitals as tested by the prevalence and spread of " erysipelas in these institutions," the same critic observed:—

" Every one knows by this time how inferior the arrangements

* "On Hospitalism," p. 10. Longmans, 1874.

" for nursing, cleanliness, and ventilation in cottage hospitals are
" to those of our great city hospitals." This last statement is made
by a gentleman who holds a deservedly high place amongst metropolitan surgeons. It is so entirely imaginary and contrary to the
fact, that I must ask him to unreservedly withdraw it. Before
doing so I should wish him to visit such hospitals as Cranleigh,
Boston, Grantham, Petersfield, Reigate, Savernake, situated as they
are in different parts of the country, and ministering as they do
to the wants of agricultural and urban populations. He will then
feel compelled to admit he has inadvertently been led to make a
charge of bad management against these crisply conducted little
hospitals which has no foundation in fact. Whatever sins may
be laid to the charge of cottage hospitals, they are certainly
not filthy, badly nursed, or ill ventilated. Taking the average,
in all these respects, the arrangements are, if anything, more
perfect than in the majority of the larger hospitals throughout
the country.

I am not disposed to quarrel with my critics for taking me to
task because I have given figures, and not a history of all the cases
contained in the tables. But in this, as in other things, it is easier
to criticise than to remedy the omissions complained of. The
labour of abstracting some 400 cases from the hospital books, of
condensing and codifying the facts, and of classifying the information so as to reduce it to reasonable but intelligible limits, is not
considered. Add to the foregoing that the facts have to be
collected from at least sixty different places scattered all over
the kingdom, and even the most exacting of critics will see
cause to be lenient in his judgment. With the view, however, of
giving information on the points referred to, I have taken out the
following facts and figures, which supply all the information
demanded by the statisticians I have quoted. Every case given in
these tables has been accurately recorded. I have full notes of the
cases in my possession, and the detailed information there given is
at the disposal of any one who may care to study it. It will be
seen that the results are more favourable to the cottage hospitals
than those given in the original tables, and that the charge of
" want of surgical boldness " (*i.e.*, refraining from amputation in
cases which would not be allowed to die in metropolitan hospitals
without amputation) is not borne out by the facts. This is credit-
able to all concerned, and adds weight to the conviction—a conviction which is spreading amongst the well-to-do classes in country
districts—that if they have to undergo an operation, it is as safe,
and on the whole more desirable to have it performed at their
own houses by the cottage hospital surgeon than to submit to the
discomforts and risks of a London lodging house, where the case

can be placed in charge of one of the more notable surgeons of a large hospital.

In this connection I have made it my business to visit many of the newly erected cottage hospitals. It is now a quarter of a century since the first cottage hospital was opened, and the older hospitals are beginning to desire to " dabble in bricks and " mortar." My observations lead me to fear that at present these new hospitals are worse for the patients than the old cottages. The former had no system of direct drainage ; the latter have a system of their own. So far as my observations have gone, I have found *the sanitary arrangements of every new cottage hospital faulty,* with one solitary exception, the Grantham District Hospital. As a matter of fact, the change from the old to the new buildings constitutes a danger to the health of the patients, for sewer gas is directly laid on to the latter, whereas earth closets or the old fashioned outside privies were probably used at the former. Architects, almost without exception, display a fatal ignorance of the most rudimentary principles of sanitary construction. Only recently a new cottage hospital was built, and the patients transferred from the old cottage, which had done good service for nearly twenty years. In this case, as usual, the closets were placed inside and in the centre of the hospital. The soil pipes were unventilated, and were directly connected with the cesspool, and many of the drains ran beneath, instead of outside, the hospital. No care in dressings, and no amount of watchfulness on the part of the medical attendant or the nurse, will prevent an outbreak of erysipelas or of something worse if the sanitary arrangements remain as I found them. The history of the new St. Thomas's Hospital and of the Leeds Infirmary proves how soon structural defects will produce septic mischief. In the new clinic of Professor Volkmann, of Halle, in Germany, though built with the utmost care, cases of cellulitis occurred within six months of the day on which it was opened. Structurally perfect, it was hygienically incomplete and unsatisfactory. Unless the cottage hospital managers set themselves steadily to work to stop this grave danger, they had best rest content with the old cottage as it is. If many fresh hospitals are built on the present bad system of construction, the mortality of cottage hospitals will, in my opinion, very soon exceed that of the larger general hospitals. Before any more new cottage hospitals are built, the staff should insist upon the plans being submitted to some competent sanitarian for his advice and counsel.

The following tables are compiled from information supplied by forty-four cottage hospitals. At very many (some sixty) others no cases of amputation had occurred.

TABLE I.—*Primary Amputations*

Name of Hospital.	Cases.	Sex and Age.	Previous State of Health.	Nature of Injury.
St. Leonard's, Sudbury	2	M. 30	Good	Compound fracture of thigh
,,	1	M. 26	Good	Railway accident. Compound fracture of leg just below knee
Crewkerne	—	—	—	—
Malvern	1	M. 39	—	Railway accident. Right leg torn off above knee; fracture of left leg; severe scalp wound
Fowey............................	1	No	particulars	—
Jarrow Memorial	3	—	—	—
		—	—	Double amputation, leg and thigh
		—	—	
Melksham	1	No	particulars	—
Grantham	3	M. 65	—	Compound comminuted; fracture of leg
		M. 19	—	Triple fracture, both bones of leg, crushing of soft parts
		—	—	
Stockton-on-Tees	1	No	particulars	—
Oswestry........................	1	,,	,,	—
South Lincolnshire	3	M. —	—	Railway accident. Compound comminuted; fracture of leg; amputation of knee
		F. —	—	Thrashing machine accident. Compound comminuted; fracture of leg, implicating knee
		No	particulars	—
St. Albans	1	,,	,,	—
Kendal	1	M. 35	—	—
Bourton-on-the-Water	1	M. 19	Healthy and of temperate habits	Leg caught and retained in a waterwheel for 20 minutes. Unchecked hæmorrhage for 2 or 3 hours
Ulverston	1	No	particulars	—
Mildenhall	1	M. 70	—	Compound fracture of leg, much hæmorrhage
Ashford	1	M. 22	Healthy	Crush, compound fracture
Cranleigh	1	M. 23	—	Crush of leg by steam thrashing machine

Remarks—Twenty-four cases. Fourteen deaths = 58·3 per cent. Shock, 9; septic resulted from shock, the most common cause of death in these cases. In all, the accidents deaths resulted from septic diseases. One patient died on the fifth day, of exhaustion, never

of Thigh for Injury.

Seat of Amputation.	Course of Case.	Result.	Cause of Death.
Thigh	—	D.	Shock
,,	—	R.	—
,,	—	D.	Pyæmia
,,	Died immediately after operation	D.	Shock
,,	—	R.	—
,,	—	R.	—
,,	—	D.	Shock
,,	—	D.	,,
,,	—	D.	,,
,,	Collapse; much hæmorrhage; lived some days	D.	,,
,,	—	D.	Septicæmia
,,	—	R.	—
,,	—	R.	—
,,	—	R.	—
,,	—	D.	Shock
,,	—	D.	,,
,,	—	R.	—
,,	—	D.	No particulars
,,	—	R.	—
Knee crushed to a pulp	Patient was collapsed, and stimulants were administered for 30 hours previous to amputation	D.	On 5th day, from shock and exhaustion
,,	—	R.	—
,,	Patient died soon after operation	D.	Shock
,,	—	D.	—
,,	—	R.	—

disease, 2; shock and exhaustion, 1; no particulars, 2. Of these fourteen deaths, nine requiring the operation were very severe, and in two there was excessive hæmorrhage. Two having recovered the shock of the accident, and in two cases the cause of death is unreturned.

TABLE II.—*Primary Amputations*

Name of Hospital.	Cases.	Sex and Age.	Previous State of Health.	Nature of Injury.
Charlwood	1	—	—	Railway accident. Compound fracture of both legs
Crewkerne	1	—	—	—
Malvern	2	M. 1½ M. 46	Good —	Compound fracture of leg Thrashing machine. Crush of leg
Bromley	1	M. 50	—	Fall of truck on leg
Dorking	3	No	particulars	—
Fowey	1	,,	,,	—
Burford, Oxon	1	M. 12	Good	Traction engine. Crushed foot
,, Tenbury	1	F. —	—	Machine. Crush of leg
Jarrow Memorial	7	M. — M. 34	— Good	— Five others without particulars
Ottery St. Mary............	2	M. 80 No	— particulars	Railway. Crush of leg
Stockton	23	All rail	way and iron	work accidents
		The	two amputati	ons fatal from shock were double
Oswestry	4	—	—	Compound fracture
Beccles	1	—	—	Double amputation
Erith	1	—	—	Amputation of foot
Bournemouth..............	1	M. 30	Navvy	Railway truck passed over both legs; comminuted fracture of front row of tarsus on left side; compound comminuted fracture of left leg; stripping of skin
St. Albans	2	—	—	—
Reigate	1	M. 50	Prematurely aged	Railway smash of leg
Ledbury	1	M. 12	—	Crushed foot and ankle
Ulverstone	1	—	—	Railway crush of foot and leg
Mildenhall	1	M. 20	—	Thrashing machine accident. Lacerated wound of leg and foot
Kendal	1	M. 26	—	Railway injury to leg
Ashford	5	M. 18 M. 26 M. 66 M. 40 M. 33	— — Healthy ,, ,,	Crush. Compound fracture ,, ,, Run over. Crushed leg Compound fracture. Tibia and fibula Crushed leg. Machinery

Remarks.—Sixty-two cases. Thirteen deaths = 20·9 per cent. Shock, 4; exhaustion, cases of death from shock, the accidents were very severe. In one, there was a compound which acute bronchitis is given as the cause of death the stump was healed, and the patient occurred in one hospital (Stockton), and the other at Ashford.

of Leg for Injury.

Seat of Amputation.	Course of Case.	Result.	Cause of Death.
—	Collapse ; died in 6 hours	D.	Shock
—	—	D.	Tetanus
—	—	R.	—
—	—	R.	—
Upper ⅓	Some sloughing, but stump good	R.	—
—	—	3 R.	—
—	—	R.	—
Lower ⅓	Perfect	R.	—
—	Shock, anæmia, bronchitis, stump healed	D.	Acute attack of bronchitis
—	Delirium tremens	D.	In 17 days, exhaustion
—	—	6 R.	—
—	—	R.	—
—	—	R.	—
—	—	16 R.	—
—	—	7 D.	{ 2 Pyæmia / 2 Shock / 3 Unreturned
—	—	4 R.	—
—	—	R.	—
Foot	—	R.	—
Right chopart, left middle, ⅓ leg	Favourable	R.	—
—	—	2 R.	—
—	Shock, no reaction, death next day	D.	Shock
—	—	R.	—
—	Did well	R.	—
Upper ⅓	—	R.	—
—	—	R.	—
—	—	D.	Pyæmia
—	—	R.	—
—	—	R.	—
—	—	R.	—
—	—	R.	—

delirium tremens, 1 ; tetanus, 1 ; acute bronchitis, 1 ; pyæmia, 3 ; cause unreturned, 3. As to the fracture of both legs, and in two others, double amputations were performed. In the case in had been accustomed to have attacks of a similar nature. Of the three cases of pyæmia, two

TABLE III.—*Primary Amputations*

Name of Hospital.	Cases.	Sex and Age.	Previous State of Health.	Nature of Injury.
St. Leonard's, Sudbury	2	M. 9	Good	Hand and forearm crushed
,,		M. 50	,,	Both arms crushed by machinery; required amputation
Crewkerne	3	No	particulars	—
Malvern	1	M. 14	—	Machinery accident. Compound luxation and comminuted fracture of humerus; fracture of femur; severe scalp wound
Bromley	2	—	—	—
,,		M. 28	—	Thrashing-machine crush
Dorking	1	M. 14	—	,,
Burford, Oxon	1	No	particulars	—
Jarrow Memorial	2	,,	,,	—
Hatfield Broad Oak	1	M. 44	Good	Avulsion of arm by rope of steam plough
Ottery St. Mary	1	M. 20	—	Avulsion of arm by thrashing-machine
Stockton	7	No	particulars	—
,,		—	—	—
Oswestry	4	—	—	Compound fractures
Hillingdon	1	M. 40	—	Machinery crush of forearm; elbow-joint implicated
Beccles	1	No	particulars	—
Newton Abbot	1	M. 25	—	Chaff-cutter accident. Crush of forearm and elbow-joint
Bournemouth	1	F.	Old and cachetic	—
Reigate	2	M. 57	Healthy	Chaff-cutter injuries to forearm and elbow
,,		M. 55	,,	,,
Bourton-on-the-Water	1	M. 16	Healthy, temperate	,,
Ledbury	2	M. 40	—	Crushed arm and elbow
,,		M. 42	—	,,
Buckhurst Hill	1	—	—	Arm severed by railway accident
Kendal	2	M. 14	—	Railway injury to arm
,,		M. 40	—	,,

Remarks—Thirty-seven cases. Five deaths = 13·5 per cent. Shock, 2; doubtful, 1; shoulder joint, and in the other there was a concomitant injury to the chest. In one case the which the cause of death is given as exhaustion, the amputation was at the shoulder joint,

of Arm for Injury.

Seat of Amputation.	Course of Case.	Result.	Cause of Death.
— Near shoulder	— —	R. R.	— —
— Shoulder joint	— Death on 10th day	3 R. D.	— Exhaustion
— Middle High — —	— Satisfactory Recovery uninterrupted — —	R. R. R. R. 2 R.	— — — — —
Shoulder-joint	—	R.	—
Close to shoulder	—	R.	—
— — — Shoulder-joint	— — Death occurred 40 minutes after operation, suddenly, while patient was conversing.	6 R. 1 D. 3 R. 1 D.	— No particulars — Entry of air in large veins (?)
—	Well in 20 days	R.	—
— Lower ⅓	— —	R. R.	— —
—	Plugged femoral vein after leaving hospital	R.	—
—	—	R. -	—
—	—	R.	—
—	—	R.	—
— —	— —	R. R.	— —
—	Injury to chest	D.	Shock, &c.
Surgical neck Shoulder-joint	— —	R. D.	— Shock

exhaustion, 1; no particulars, 1. In the two deaths from shock, one amputation was at the patient died suddenly after the operation, the cause of death being doubtful; in the case in and there was also a fracture of femur and a scalp wound due to the same accident.

Table IV.—*Amputations*

Name of Hospital.	Cases.	Sex and Age.	Previous State of Health.	Nature of Injury.
St. Leonard's, Sudbury	2	M. 53 M. 12	Good ,,	Hand torn off by machinery Hand crushed ,,
Crewkerne	2	No	particulars	—
Bromley	1	M. —	—	Avulsion of hand by rollers of chaff-cutter
Dorking	1	No	particulars	—
Burford, Tenbury	2	M. — No	— particulars	Steam saw accident —
Jarrow Memorial	1	M. 28	Good	—
Melksham	1	No	particulars	—
Hatfield Broad Oak	1	M. 28	Good	Shattering of bones of wrist, &c., by chaff-cutter
Ottery St. Mary............	2	M. 7 F. 17	— —	Machine crush of hand ,,
Grantham	5	No	particulars	—
Stockton	5	,,	,,	—
Oswestry	2	,,	,,	—
Louth	1	,,	,,	—
Beccles	1	,,	,,	—
St. Albans	3	,,	,,	—
Reigate	1	M. 25	Good	Chaff-cutter smash of hand
Bourton-on-the-Water	1	M. 16	Good health; temperate	Compound fracture
Ulverston	2	M. 22 M. 38	— —	Crushed hand ,,
Ashford	2	M. 13 M. 18	Good health Good, but much exhausted by hæmorrhage	Machine crush Gunshot injury to bones and arteries
Cranleigh	1	M. 25	Navvy	Compound comminuted fracture, with dislocation of left wrist; compound comminuted fracture of left ankle; laceration of lung
Kendal	1	—	—	—

Remarks.—Thirty-eight cases. One death = 2·6 per cent. The only death in this table in the same accident. At the time of death the amputation was progressing favourably.

f Forearm for Injury.

Seat of Amputation.	Course of Case.	Result.	Cause of Death.
—	—	R.	—
—	—	R.	—
—	—	2 R.	—
Just above wrist	Satisfactory	R.	—
—	—	R.	—
Lower ⅓	Small abscess	R.	—
—	—	R.	—
—	—	R.	—
—	—	R.	—
Lower ⅓	—	R.	—
—	—	R.	—
—	—	R.	—
—	—	5 R.	—
—	—	5 R.	—
—	—	2 R.	—
—	—	R.	—
—	—	R.	—
—	—	3 R.	—
—	Satisfactory	R.	—
—	—	R.	—
—	—	R.	—
—	—	R.	—
—	—	R.	—
—	—	R.	—
—	Effusion of blood beneath sternum : died from effect of injuries on 11th day. Amputation progressing favourably	D.	Concomitant wound of lung, &c., &c.
—	—	R.	—

was in a patient who had sustained a compound fracture of ankle and a wound of the lung

TABLE V.—*Secondary*

Name of Hospital.	Cases.	Sex and Age.	Previous State of Health.	Nature of Injury.
Bromley........................	1	—	Stout, pale, flabby, a notorious drinker	—
Oswestry	1	M. 40	Delicate constitution	Fracture of leg
Burford, Tenbury........	1	M. —	Unhealthy	Simple fracture lower half of tibia, upper one-third fibula
Ulverston	1	M. 23	Phthisical subject	Compound fracture of thigh
LEG.				
Ottery St. Mary	1	M. 50	—	Compound fracture of tibia and fibula from upsetting of waggon, uncontrollable hæmorrhage for many days
ARM.				
Bourton.......................	1	M. 17	Irregular habits	Compound fracture of humerus, caused by bursting of steam engine
FOREARM.				
Newton Abbot	1	M. 30	Good health	Gunshot wound of hand; thumb, forefinger, and metacarpal bones of two middle fingers removed same day
St. Leonard's, Sudbury	1	M. 46	Always good	Part of hand crushed by machinery, one finger amputated
LEG.				
Erith	1	M. 50	—	Compound fracture of tibia and fibula, implicated ankle, and much laceration of soft parts

Remarks.—Nine cases. Three deaths = 33·3 per cent. Of the two deaths in secondary exhaustion, in a stout, pale, feeble man—a notorious drinker, the subject of traumatic was gangrene, following uncontrollable hæmorrhage, lasting many days.

Amputations for Injury.

Nature of Disease.	Seat of Amputation.	Course of Case.	Result.	Cause of Death.
Traumatic gangrene	Thigh	—	D.	Shock
Gangrene of leg	Thigh	Healed rapidly; about in 14 days	R.	—
Phlegmonous erysipelas, sloughing of intermuscular septa as high as knee	Thigh, Lower ⅓	Slow repeated attacks of second hæmorrhage; discharged in five months	R.	—
Amputated many weeks after	Thigh	Attack of pneumonia	D.	Pneumonia
—	Leg	Gangrene of the stump occurred	D.	Gangrene
Gangrene of whole arm appeared on fourth day	Shoulder	Some sloughing, but patient made a good recovery	R.	—
Acute cellulitis extending two inches above wrist joint; amputated on 14th day	Middle ⅓	—	R.	—
Abscesses and inflammation	Lower ⅓ forearm	—	R.	—
Surgical fever inflammation	Upper ⅓ of leg	Satisfactory	R.	—

amputation of thigh, one was from pneumonia in a phthisical subject, and the other from gangrene. In the one fatal case following secondary amputation of leg, the cause of death

TABLE VI.—*Amputations*

Name of Hospital.	Cases.	Sex and Age.	Previous State of Health.	Nature of Disease.
St. Leonard's, Sudbury	2	M. 12	For years a succession of strumous abscesses	Necrosis of tibia extending to knee
"	—	M. 74	Good	Ununited fracture of tibia and fibula
Malvern	8	F. 36	"	Strumous disease of knee joint
"	—	F. 24	Fair	"
"	—	M. 5	Debilitated subject	"
"	—	M. 31	Health much broken	Strumous disease of knee, old
"	—	F. 23	Good	Strumous disease of knee
"	—	F. 53	Fair	Gelatinous disease of synovial membrane of knee
"	—	M. 19	Very reduced and emaciated	Strumous disease of knee, old
"	—	M. 14	Good	Strumous disease of knee
Enfield	1	M. 7	Delicate, badly nourished	Abscess, necrosis of internal condyle of femur, disorganisation of knee-joint
Dorking	1	M. 10	Chronic eczema	Necrosis of tibia, implication of knee-joint
Hatfield Broad Oak	1	M. 37	Very emaciated, bad health	Osteo-sarcoma of femur two years ; refused earlier operation
Ottery St. Mary	2	Ch. 11	Debilitated	Necrosis of tibia, profuse discharge
"	—	F. 59	—	Anchylosis of knee, and constantly recurring abscesses in calf
Grantham	1	No	particulars	—
Stockton	2	"	"	—
"	—	—	—	—
South Lincolnshire	3	No	particulars	—

of Thigh for Disease.

Seat of Amputation.	Course of Case.	Result.	Cause of Death.
—	—	R.	—
—	—	R.	—
—	—	R.	—
—	—	R.	—
—	—	R.	—
—	—	R.	—
—	—	R.	—
—	—	R.	—
—	Patient died on 5th day	D.	Exhaustion and diarrhœa
—	—	R.	—
Amputation of middle and lower ⅓ disartic hip	Continuation of necrosis until disarticulation, erysipelas of head	R.	—
—	—	R.	—
—	Secondary hæmorrhage 14 days after operation, ligature of common femoral, recurrence of hæmorrhage in 16 days, ligature of external iliac, died of hæmorrhage in 14 days more	D.	Secondary hæmorrhage
—	—	R.	—
—	—	R.	—
—	—	R.	—
—	—	R.	—
—	—	D.	—
—	—	3 R.	—

TABLE VI.--*Amputations of*

Name of Hospital.	Cases.	Sex and Age.	Previous State of Health.	Nature of Disease.
Hillingdon	1	F. 17	Bad, since excision twelve months previous	Sinuses, no bony union, unpromising excision of knee
Erith	1	M. 9	Scrofulous, delicate	Acute periostitis, knee-joint implicated
Reigate	1	M. 23	Strumous	Strumous disease of knee, previous excision
St. Alban's	1	—	—	—
Ledbury	1	F. 38	—	Medullary cancer of knee
Ulverstone	2	F. 4	—	Medullary cancer of lower end of femur
,,	—	F. 44	—	Chronic disease of knee-joint
Mildenhall	1	M. 50	Great emaciation	Synovial disease of knee-joint
Chesham	1	M. 21	Good previous to development of cancer	Cancerous tumour of internal condyle of femur
Ashford	2	F. 39	Very cachectic	Fungus hæmatodes
,,	—	M. 36	Very unhealthy, strumous	Disease of knee-joint
Cranleigh	4	M. 10	—	Scrofulous disease of knee
,,	—	M. 15	Very reduced and feeble	,, and femur
,,	—	M. 26	External strumous, anæmia	,, 6 months
,,	—	M. 9	—	Carcinoma of femur
Kendal	4	M. 36	—	Disease of knee
,,	—	M. 10	—	Ill 15 months, disease of knee
,,	—	M. 2	—	Disease of knee
,,	—	M. 14	Ill two years	,,

Remarks.—Forty cases. Six deaths = 15 per cent. Exhaustion, 3 ; secondary hæmorr-disease, five of which recovered. "Many of the amputations for knee disease would probably following amputation for malignant disease was from exhaustion following secondary hæmorr-refused an earlier operation.

Thigh for Disease—Contd.

Seat of Amputation.	Course of Case.	Result.	Cause of Death.
—	Good	R.	—
—	Amputation of other leg, good	R.	—
—	Secondary hæmorrhage on third day	D.	Exhaustion, secondary hæmorrhage
—	—	R.	—
—	—	R.	—
—	—	R.	—
—	—	R.	—
—	—	R.	—
—	—	R.	—
—	—	R.	—
—	—	R.	—
—	—	R.	—
—	—	R.	—
—	Died in 6 weeks	D.	Exhaustion.
Below trochanter	—	R.	Living many years after.
—	—	R.	—
Lower ⅓, afterwards higher	—	D.	Exhaustion, suppuration.
—	—	R.	—
—	—	R.	—

hage and exhaustion, 2; no particulars, 1. This table contains six amputations for malignant not have been done in London, where resection is more common." Of the deaths, the one hage, after successive ligature of the common femoral and external iliac arteries; this patient

TABLE VII.—*Amputations*

Name of Hospital.	Cases.	Sex and Age.	Previous State of Health.	Nature of Disease.
St.Leonard's, Sudbury	1	F. 57	Always in bad health	Extensive caries of tarsus
Market Rasen	1	M. 40	Thin, anæmic	Periostitis, followed by necrosis of tibia. Spontaneous fracture
Malvern	3	M. 65	—	Recurrent necrosis of tarsal and metatarsal bones
	—	M. 22	Good health	Painful pirogoff, stump
	—	—	—	—
Fowey	3	No	particulars	—
Ottery St. Mary	1	M. 78	—	Senile gangrene
Stockton	2	No	particulars	—
Erith	1	M. 9	Scrofulous, delicate	Acute periostitis of tibia. One knee and one ankle implicated.
Reigate	2	M. 12	Strumous	Necrosis of tibia, and suppuration of ankle, following wound of joint
	1	M. 55	—	Compound fracture into ankle. Necrosis of tibia
Cranleigh	1	M. —	—	Periostitis, resulting in disorganisation of right foot
Kendal	3	F. 9	—	Disease of ankle-joint
	—	M. 6	—	,,
	—	F. 60	—	Disease of ankle and ulcer of leg
Erith	1 ft.	M. 76	Declining strength	Senile gangrene of great toe

Remarks.—Nineteen cases. Three deaths = 15·7. In one case no particulars of the the patient recovered from it and lived five weeks, then dying suddenly.

of Leg for Disease.

Seat of Amputation.	Course of Case.	Result.	Cause of Death.
—	Died four years after of cancer of liver	R.	—
—	Up and poaching in one month	R.	—
Since amputation of great toe, foot, and lower ⅓ leg	—	R.	—
Lower ½	—	R.	—
—	—	R.	—
—	—	3 R.	—
—	Operation recovered from. No further gangrene	—	Died suddenly five weeks after operation
—	—	{ R. D. }	No particulars
Upper ⅓	Other thigh amputated. Good recovery	R.	—
—	Good	R.	—
—	Sharp attack of idiopathic erysipelas	R.	—
Upper ⅓	Good	R.	—
Lower ⅓ scale	—	R.	—
„	—	R.	—
Upper ⅓	—	R.	—
Chopart	—	D.	Exhaustion

cause of death are given. In the first case the operation might claim to be successful, since

TABLE VIII.—*Amputations of Arm for Disease.*

Name of Hospital.	Cases	Sex and Age.	Previous State of Health.	Nature of Disease.	Seat of Amputation.	Course of Case.	Result.	Cause of Death.
Milton Abbas	1	F. 40	Always delicate. Health much undermined	Aneurismal varix following wound over wrist, gangrene of two fingers. Veins much enlarged up to shoulder	—	Good	R.	—
Jarrow	2	No particulars		—	—	2 R.	—
Ottery St. Mary	1	"	"	—	—	R.	—
Grantham	1	"	"	—	—	R.	—
Hillingdon	1	M. 10	Hectic, emaciated	Elbow disease. Excision two months previously	—	Good	R.	—
Mildenhall	1	M. 17	—	Synovial disease of elbow	—	—	R.	—
Cranleigh	1	F. 23	Feeble. Seven months pregnant	Scrofulous "	—	Good. Child born alive at full time	R.	—

Remarks.—Eight cases. No deaths.

TABLE IX.—*Amputations of Forearm for Disease.*

Name of Hospital.	Cases.	Sex and Age.	Previous State of Health.	Nature of Disease.	Seat of Amputation.	Course of Case.	Result.	Cause of Death.
Bromley	1	F. 70	—	Necrosis of carpus. Suppuration. numerous fistulæ	Upper $\frac{1}{3}$	Satisfactory	R.	—
Bourton	1	F. 17	—	Malignant disease of hand and wrist	—	—	R.	—
Ledbury	1	M. 65	—	Strumous disease of hand and wrist	—	—	R.	—
Cranleigh	1	M. 77	—	Epithelioma of back of hand	—	—	R.	—

Remarks.—Four cases. No deaths.

General Summary of all the Cases.

Results of Amputation. Cottage Hospitals. | **Results of Amputation. University College Hospital.**

Seat.	Cases.	Recoveries.	Deaths.	Percentage of Deaths.			Seat.	Cases.	Recoveries.	Deaths.	Percentage of Deaths.
Thigh	24	10	14	58·3			Thigh	39	16	23	59·0
Leg and foot	62	49	13	20·9	Amputation for Injury.		Leg and foot	44	30	14	31·8
Arm	37	32	5	13·5			Arm	12	7	5	41·6
Forearm	38	37	1	2·6			Forearm	8	8	0	0·
Totals	161	128	33	20·4			Totals	103	61	42	40·7
Thigh	40	34	6	15·0			Thigh	86	68	18	20·9
Leg and foot	19	16	3	15·7	Amputation for Disease.		Leg and foot	74	64	10	13·5
Arm	8	8	—	—			Arm	24	16	8	33·3
Forearm	4	4	—	—			Forearm	20	19	1	5·0
Totals	71	62	9	12·6			Totals	204	167	37	18·1

Secondary Amputations for Injury in Cottage Hospitals.

Thigh	4	2	2	9 cases, mortality 33·3 per cent.
Leg	2	1	1	
Arm	1	1	—	
Forearm	2	2	—	
Totals	9	6	3	33·3

Total Numbers in Town and Cottage.

	Cases.	Deaths.	Percentage of Mortality.
Cottage Hospitals	241	45	18·6
University College	307	79	25·7

With regard to the occurrence of septic disease, the statistics given in the above tables are very favourable to cottage hospitals. Mr. Bryant states that in Guy's Hospital 10 per cent. of all amputations die from pyæmia, and that 42 per cent. of the fatal cases may be traced to this cause. Now, on examining these 241 cases, we find 5 cases of septic disease, 4 of pyæmia (2 of these occurring in one hospital), and 1 case of septicæmia, against a total of 45 deaths; so that the percentage of deaths from septic disease to the total number of cases reaches only 2·1, and the percentage of deaths from this cause to the fatal cases is only 11·1. These cases of septic disease all occurred after amputations of the lower extremity. In no case of amputation for disease did pyæmia or septicæmia occur. Of the 2 deaths in the 8 cases of secondary amputation for injury, neither was due to septic poisoning. (In Mr. Erichsen's cases, as many cases of pyæmia occurred after amputation for disease as in primary amputations for injury.) May not this fact be taken as conclusive evidence in favour of the healthiness of small as compared with large hospitals?

The facts contained in the foregoing tables reflect upon the justice of the assertion that there is greater surgical boldness displayed by the London surgeons, as some of the operations point to a surgical skill and boldness which leave nothing to be desired. For instance, a case of amputation of thigh was performed in the Hatfield Hospital for malignant disease of femur, in which both the common femoral and the external iliac arteries were successively ligatured for secondary hæmorrhage; and an amputation of thigh followed by exarticulation of hip was successfully carried out at Enfield.

Again, as to the undertaking of operations, the amputations of thigh for ununited fracture of leg in a patient of 74, successfully performed at St. Leonard's, Sudbury, and that for senile gangrene planned and successfully carried out at Ottery St. Mary, are favourable specimens of surgical boldness combined with judgment. The case of amputation of arm at Milton Abbas, performed after the case had been rejected at the County Hospital, also speaks well for the surgical staff there.

On this subject the following letters from three cottage hospital surgeons will prove of interest:—

Mr. Thomas Moore, F.R.C.S., San. Sc. Certf. Cantab., writes:—

"As regards the comparative influence of the air of large and small hospitals on the healing of wounds, about which you ask my opinion, I do not hesitate to say that I believe they do better as a rule in the smaller institutions.

"I have unfortunately not had the immense experience of operations which falls to the lot of some surgeons in large towns, but I have had considerable opportunities of studying the vexed question on both sides. First, for five years (during eighteen months of which time I was surgeon's dresser) at St. Bartholomew's Hospital, and for one year at the Queen's Hospital at Birmingham; then as surgeon for six years to two of the largest ironworks in Staffordshire; and lastly, for between eight and nine years as surgeon and hon. secretary to the Petersfield Cottage Hospital, and in a large private practice.

"*Cæteris paribus*, I found that the very numerous cases of compound fracture and severe wounds I was called upon to treat in the ironworks and their attached collieries did better, as a rule, when treated isolated in the patients' own cottages than when removed to a hospital, in spite of inferior nursing and poor feeding. That was, however, I am bound to say, when I looked after them daily or more frequently, *myself*; for the old saying, 'cleanliness is next to godliness' is more than true when applied to surgical dressings.

"The way in which wounds heal in the pure air of this cottage hospital (Petersfield), where there are good nursing, and every available creature comfort, and where no ward contains more than two beds, is beautiful; and I think I could persuade even Mr. Lister himself that antiseptic precautions are not necessary under all circumstances, if I could get him to spend twelve months in the unexciting but germless atmosphere of our cottage hospital. Out of 272 surgical cases, many of them of a serious nature, only one has been attacked by erysipelas, and that occurred when there were several cases of puerperal fever in the neighbourhood, and could thus probably be accounted for. It is only right to mention that some of the healthiness of this hospital may be attributed to the fact that we have no sewer gas laid on, as is too frequently the case, I fear, even in the best drained towns. The closets are on the earth principle, and the drain from the kitchen sink is made to open well into the outer air.

" It has been urged against the statistics which have been brought forward to prove the superior suitability of cottage hospitals for the treatment of severe surgical cases, that the more severe ones are rejected in them, and are sent to the larger hospitals. This has certainly not been the case here. Two or three cases of a chronic nature, where the advisability of an operation was doubtful, have been sent away, but every case of accident has been taken in without inquiry, and some were of a most severe and unpromising nature. On the other hand, it is well known to the initiated that some of the great London operators are very careful to select their cases, and this they can easily do without its being known ; whereas if a country surgeon declines to operate, all the neighbourhood likes to know the ' why and wherefore,' and is apt to make invidious comparisons.

" It has been urged against the establishment of cottage hospitals that the same amount of surgical skill cannot be brought to bear upon the cases as in larger ones. That may be, and it would savour of egotism on my part to deny it. Still if statistics prove that cases (of amputation for example) get on better in the former than in the latter, is it not better for the patients to have pure air and less skill brought to bear upon their ailments ?

" Moreover, this state of things will tend to mend itself year by year, as severe cases are more and more treated by the local surgeon, and are not sent off to the county infirmary, as was formerly the rule. I lose no opportunity of urging on the numerous rich residents of this neighbourhood the fact, that in supporting the cottage hospital, they are but ' casting their bread upon the waters,' for the more practice the local medical men get in bad accidents and operations among the poor people, the better will they be able to treat emergencies among the rich, and the less necessity will there be for the ' eminent consultant ' and his 50 guineas' fee, and they are, I believe, beginning to see the truth of the remark.

" A well known gentleman, and an enthusiastic supporter of medical institutions, argued with me a short time ago against the institution of cottage hospitals, and, as a kind of ' clencher,' averred that they are damaging the county infirmaries. I fear that may be so, but in this matter the ' greatest good of the greatest number ' must be considered. The latter have had their day, and *have* done much good, but if the former are calculated to do more still, surely no sentimental idea of that kind should be allowed to stand in their way. The larger institutions will still be very useful to receive the chronic and incurable cases, even if they do not attract the accidents and operations so much."

Mr. Thomas H. Cheatle, of Burford, writes :—

" As to the necessity for an operation, the case and the common sense of the surgeon determine the question. Of course in any difficult case further advice and assistance would be obtained. In a purely agricultural district like this, with little machinery, and the people becoming more used to what there is, there is little surgery in the way of operations to be had, and it is quite possible that the country surgeon, while he is careful to avoid temerity, may seem to lack the ' boldness ' which is assumed to be the characteristic of the urban operator."

Mr. W. Berkeley Murray, of Tenbury writes :—

" My general rule has always been to ask the advice of my colleagues, and upon a conviction that an attempt to save the limb would be attended by danger to life of the patient, I have operated without unnecessary delay. I think there can be no doubt that there is greater surgical boldness shown by surgeons of large hospitals, and reasons for this are not far to seek. The authority of the man with a name, and the authority of the large and old established institution cover all ill success, and it is responsibility which produces caution, not to mention the boldness given by constant practice. Nevertheless, we undoubtedly possess *great* advantages in the pure country air and quiet, and in the concentrated care we are able to bestow on any bad case."

The above remarks are forcible and convincing, and they will certainly carry weight.

Two points may be referred to in this connection. It must not
be overlooked in considering the question of surgical boldness, that
a surgeon to a large clinical hospital is under the necessity of
remembering that he has as far as possible to cure the greatest
number of patients in the shortest possible time. Hence a surgeon
so situated is under the necessity of operating frequently because
of the crowded state of the hospital and the great demands upon
its available space. Such circumstances render speedy results an
absolute necessity.

Again, the long distance which patients have to be carried to
reach the cottage hospital, as compared with that traversed by
accident cases in large towns, may reasonably be considered to
increase the deaths from shock, and to add to the severity of the
conditions which render recovery improbable. In large towns not
only is the distance shorter, but the patients are more accustomed
to think at once of the hospital, and there conveyances are always
to be had. These are, therefore, not unimportant considerations.

It has taken some years to compile and complete the statistics
contained in this paper. Early in 1876 I commenced to collect the
first portions of these statistics, which were given in my book on
cottage hospitals,* published in 1877. In consequence of the criti-
cism which these tables elicited, I resolved to still further investi-
gate the subject, and after nearly two years' labour I obtained the
history of each of the two hundred and forty-one cases with which
I have been dealing in the latter part of this paper. It will therefore be observed that the compilation of my statistics has occupied
nearly four years, and that I was not able to complete my second
set of tables, which include a history of each case enumerated, until
the end of the year 1880. I then offered to read a paper before this
Society. The proposal was cordially entertained by the Council ;
but owing to various circumstances, it has not been possible to
present it for the consideration of the Fellows until this evening.
Thus nearly six years have elapsed since I first commenced my
investigations on this instructive subject.

Science never stands still : all true science is progressive. There
is no finality of which we can be certain where science is concerned. It follows almost as a matter of course that this period of
six years has revolutionised the treatment of cases of operation
and open wounds, both in hospital and private practice. Aseptic
surgery since the year 1878 has been making rapid and convincing
progress. Its results have practically cut the ground from under
the feet of those who with anxious care formerly debated the
question of the relative mortality of large and small hospitals. I

* "The Cottage Hospital; its Origin, Progress, Management, and Work,"
first edition, 300 pp. London : J. and A. Churchill.

am therefore, as an honest searcher after scientific truth, induced to-night to declare that the aseptic system of Mr. Lister has practically solved this great question, by proving that where this method of treatment is carefully enforced, the size and condition of the hospital buildings is of comparative unimportance. It would be wrong of me to content myself with a bare statement of this important fact, and I therefore proceed to give the evidence upon which my declaration is founded. To enable me to do this I must trouble you once again with statistical tables. These tables contain an account of the results obtained in Germany from two hundred and thirty-four cases of amputation from various causes, all of which were treated aseptically on Professor Lister's plan. It has been necessary to somewhat alter my statistics, so that they may exactly correspond with those prepared by the eminent German surgeon Dr. Schede, of Hamburgh.* This has slightly reduced the number of cases given in my tables, because Dr. Schede omits double amputations, cases in which other severe injuries co-exist, and cases in which intercurrent diseases not related to the operation carry off patients whose stumps are healed. This has slightly improved the percentage in cottage hospital practice. In spite of this, however, the mortality in cottage hospital practice stands at 15·3 per cent., as against 2·9 per cent. in the cases recorded by Dr. Schede, which were treated in large German hospitals, on the Listerian or aseptic system. It thus follows that whereas Sir James Simpson gives an average mortality in town hospitals of 41·6 per cent., and although Mr. Erichsen was proud to be able to prove some fifteen years ago that the average mortality from all the amputations performed in the wards of University College Hospital from its foundation, a period of thirty-eight years, was only 25·7 per cent., the late Mr. Callender, in his papers in the St. Bartholomew's Hospital Reports, 1869, p. 263, showed that the mortality after amputation in certain country hospitals was 17·5 per cent.; and I have shown that the results attained in cottage hospital practice give a mortality of but 15·3 per cent. Dr. Schede proves beyond dispute that Mr. Lister, by his wonderful discovery, has enabled the surgeons who adopt it conscientiously, irrespective of the size of the hospital buildings, to reduce the mortality in such cases to 4·36 per cent.† The following tables, which have been prepared by my friend Mr. G. H. Makins, who has rendered me much valuable assistance in the compilation of this paper, give Dr. Schede's and my own figures in detail :—

---

* *Vide* Dr. Schede's article in Pitha and Billroth's "Handbook of Surgery."

† The actual mortality on all Dr. Schede's collected amputations. In the tables given above, the amputations at the hip and shoulder joints are excluded, as no such cases occurred in the cottage hospital practice.

Amputations for Injury.

Seat.	Schede's Antiseptic Statistics.				Burdett's Cottage Hospital Statistics.			
	Cases.	Reco-veries.	Deaths.	Per-centage of Deaths.	Cases.	Reco-veries.	Deaths.	Per-centage of Deaths.
Thigh	23	18	5	21·7	22	10	12	54·5
Leg	19	19	—	—	57	48	9	15·7
Arm	20	20	—	—	35	31	4	11·4
Forearm	34	34	—	—	37	37	—	—
Total	96	91	5	5·2	151	126	25	16·5
Complicated Cases—								
Double amputations	13	10	3	23·0	5	1	4	80·0
Severe multiple injuries	11	3	8	72·7	4	—	4	100·0
Deaths from intercurrent disease	27	11	16	59·6	1	—	1	100·0
Total	51	24	27	52·9	10	1	9	90·0

Amputations for Disease.

Seat.	Schede's Antiseptic Statistics.				Burdett's Cottage Hospital Statistics.			
	Cases.	Reco-veries.	Deaths.	Per-centage of Deaths.	Cases.	Reco-veries.	Deaths.	Per-centage of Deaths.
Thigh	63	62	1	1·5	40	34	6	15·0
Leg	50	49	1	1·0	19	16	3	15·7
Arm	12	12	—	—	8	8	—	—
Forearm	13	13	—	—	4	4	—	—
Total	138	136	2	1·4	71	62	9	12·6
Complicated Cases—								
Double amputation	—	—	—	—	—	—	—	—
Severe multiple injuries	—	—	—	—	—	—	—	—
Deaths from intercurrent disease	—	—	—	—	—	—	—	—
Total	—	—	—	—	—	—	—	—

All Uncomplicated Amputations for Disease or Injury.

Seat.	Schede's Antiseptic Statistics.				Burdett's Cottage Hospital Statistics.			
	Number of Cases.	Recoveries.	Deaths.	Per-centage of Deaths.	Number of Cases.	Recoveries.	Deaths.	Per-centage of Deaths.
Thigh	86	80	6	7·0	62	44	18	29·0
Leg	69	68	1	1·4	76	64	12	15·7
Arm	32	32	—	—	43	39	4	9·3
Forearm	47	47	—	—	41	41	—	—
Total	234	227	7	2·9	222	188	34	15·3

Callender, in his papers in the Bartholomew Hospital Reports, 1869, p. 263, shows that the mortality after amputation in country hospitals was 17·5 per cent. (and that in old days).

        Country patients in St. Bartholomew's Hospital .... 17·0 per cent.
            „   private patients in London .... 17·1   „
            „   „   country .... 10·8   „

The figures and tables here given show fairly and truthfully what has been the saving of life owing to the adoption of the aseptic or Listerian treatment of wounds ; that is to say, a mortality of 41·6 per cent. in 1868, and a mortality in 1872, according to Professor Erichsen, of 37·8 per cent. in the larger metropolitan hospitals, and of 25·7 per cent. in University College Hospital, has fallen in Germany, under the Listerian method, to 4·36 per cent. in 1880. These figures are so remarkable as to be almost incredible. The results obtained in different classes of operations have been equally noteworthy. Thus Mr. Spencer Wells, the eminent ovariotomist, by whose direct agency Lord Selborne,* in a public address, once demonstrated 22,272 years of human life may be estimated as having been added to society, gave† the result of the last 168 cases which he had treated in private practice as follows :—The first 84 had been treated by the old methods, "but "yet with all the care I could give to them, there were 21 deaths." The last 84 were treated aseptically, and of these only 6 died. Nor is this all, for he adds: "As I went on and became still more "accustomed to the method and details of antiseptic treatment, "and avoided mistakes, then I obtained the long run of 38 cases "without a single death." Could anything more gratifying, or more honourable to a great profession, be imagined than the fact that the operation of ovariotomy has been performed in 38 consecutive cases without a single death, when it is remembered that this operation (ovariotomy), so recently as the year 1857,‡ was in "absolute disrepute." I have shown how many hundreds of lives have been saved, and are being saved, every year by the aseptic, or Mr. Lister's system. I have produced evidence that this new departure has rendered it a matter of secondary importance whether serious cases of injury and operations are treated in large or in small hospitals. I have proved that at least twenty-two thousand years of health, of usefulness, and happiness, have been added to the life of woman in Great Britain by the direct agency of another eminent surgeon, Mr. Spencer Wells. Yet to the honour of these gentlemen, but to the dishonour of our nation, neither have received any public recognition at the hands of those who distribute the national honours and rewards in this country. How has it happened that two such men as Lister and Wells, whose names are no small glory to England as benefactors to humanity at large, have not ere this received the highest honour which Government ever bestows upon medical men ? Is it because fashion or custom, or both, have decreed that Courts and Governments should confer the highest honours

---

* "British Medical Journal," 1880, vol. i, p. 932.
† Debate on Antiseptic Surgery, December, 1879.
‡ "British Medical Journal," 1880, vol. i, p. 931.

on those who are most successful in destroying life on a large scale, and not on those who save life? At any rate, be the cause what it may, the fact remains; and a damaging fact it is. No wonder if the flower of our university youth choose the Church, or the law, the army or navy, or some branch of the civil service of the State, rather than the medical profession, because they at once take an enviable social position, and a successful career may lead to titles and pensions, and to a seat in the House of Lords. It has been well asked, why should a baronetcy be the highest titular distinction conferred upon members of the medical profession? Is Jenner or Paget less worthy of a life peerage than the eminent men who now sit on the bench of bishops, or any of the lawyers, soldiers, or sailors who have been rewarded by hereditary peerages? Can any member of the House of Lords do greater service to his country in that assembly than an eminent member of the medical profession could render in the promotion of legislation for securing and protecting the public health? I for one think not. Yet in spite of the enormous saving of life effected under providence by the direct labours of Lister and Wells, neither have yet received a shade of a shadow of recognition from the Government of this country. Such an anomaly should not long continue. It is time public attention was called to it, for then those who save life will share the honours, the rewards, and the pensions with those who destroy life. Then, and not till then, justice will be tardily done to the members of a great profession, whose services are rendered to each and all of us at the time of our sorest need and greatest suffering. The sooner this awakening of the public conscience takes place, the better will it be for the national credit. Such at any rate is my view of the case, and as one aspect of the great question I have been considering in this paper, I hope this expression of opinion may not be inopportune or without practical results.

APPENDIX A.

" SIR,

"The relative success of the graver operations in surgery as
"performed, first, in large town hospitals, and, secondly in country
"cottage hospitals, has for years attracted much attention, and
"there is reason to believe that the mortality in cottage hospitals in
"the major operations is much less than in the London hospitals.
"With a view of setting this question at rest, and of proving the
"truth or fallacy of Sir James Simpson's theory, I shall feel
"deeply obliged if you will fill up the enclosed form with the
"results of all the amputations which you may have had in connec-
"tion with your cottage hospital since it was first opened.

"However few may be the amputations of the limbs, an exact
"return from every cottage hospital will be regarded as a very
"valuable contribution to surgical statistics.

"Yours faithfully,

"HENRY C. BURDETT.

" *To Dr.* ————,

" *Medical Officer of* ——————— *Cottage Hospital.*"

Result of Amputation of the Limbs in Cottage Hospital Practice.

Return from _____Cottage Hospital, having_____Beds.

Seat of Amputations.	Primary, or for Injury.		Secondary, or for Disease.	
	Number of Cases.	Number of Deaths.	Number of Cases.	Number of Deaths.
Amputation of thigh....				
,, leg........				
,, arm				
,, forearm				
Total				

Signature_____

Residence_____

Date_____

REMARKS.

Note.—The cause of each death should be noted if possible, viz., whether
secondary hæmorrhage, shock, pyæmia, or other cause. A short history of each
case should also be given.

DISCUSSION *on* MR. BURDETT'S PAPER.

MR. ROBERT LAWSON said that anyone who was accustomed to such work, would understand that Mr. Burdett's paper could not have been compiled without a great deal of assiduous attention and hard work, and they were therefore much indebted to him for the trouble he had taken. The subject of cottage hospitals had been gradually taking an increased hold upon the country. Long ago Sir James Simpson pointed out the advantages to be derived from such places, but his results were objected to by several of the surgeons of large hospitals. With perhaps an allowable feeling of self-esteem, those surgeons thought that the results in their hospitals could not be exceeded, and that if apparently better results had been obtained elsewhere, it must have been owing to the fact that less serious cases had been dealt with. Some people fancied that the statistics of disease should be taken, instead of injuries, to test this question; but any one acquainted with the phases of disease knew that such a task was attended with very great difficulty. There were such varieties of nomenclature, that it was difficult to say when a case of inflammation of the lungs, for instance, began and when it ceased; but when dealing with a limb torn by machinery, or a bone fractured, they had something tangible before them, and could draw their own conclusions. If the results in such cases in cottage hospitals were more favourable than in large hospitals, it was natural to conclude that the same advantages would be obtained in the treatment of ordinary diseases. The author showed there were great differences in the mortality following operations on different parts of the limbs. Amputation of the thigh was attended with much more danger to life than amputation of the leg; amputation of the leg than amputation of the arm; and amputation of the arm than amputation of the forearm. In calculating the risk attending any single series of operations on one of those parts, it might be done as in the table given, where there were twenty-four cases of amputation of the thigh in cottage hospitals—ten recoveries and fourteen deaths; but it would not do to take a group, as the author had done, embracing injuries of the thigh, leg, arm, and forearm, and compare them with another group in University College Hospital, where those injuries occurred in different proportions. For instance, the injuries for which amputation was necessary of the thigh and leg and foot in the cottage hospitals were 53 per cent. of the total, while those of the arm and forearm were 47 per cent. In the paper that group was compared with Mr. Erichsen's results at the University College Hospital, where the amputations of the thigh and leg amounted to 81 per cent., and of the arm and forearm to only 19 per cent. It was clear that those two groups could not statistically be compared with one another. He (Mr. Lawson) had taken the 161 cases mentioned of primary injury and divided them according to the same percentage as Mr. Erichsen's figures, and he had calculated the mortality, using Mr. Burdett's own ratio of mortality in the cottage hospitals. The result was as follows: of the 161 injuries, 52·9 would have died; that would be a mortality of 32·8 instead of 20·4. There was then still a marked difference of 7·9 per cent. in favour of the cottage hospitals over the University College Hospital. Then the question arose, was that difference accounted for by the cottage hospitals receiving a less serious class of cases, or by something in the hospitals themselves? Every practical man would allow that large hospitals had the worst class of cases as regards constitution to deal with; but that was not sufficient to account for it. The antiseptic treatment had reduced the mortality from 40·7 to an extremely small figure, but the antiseptic dressing would only affect some elements of the case. It would not affect "shock," though it would have a considerable influence upon those causes which led to pyæmia or septicæmia, one of which was the absorption of pus and the other the absorption of putrid matter, which poisoned the system. In the cases mentioned in the paper the aseptic treatment had reduced the mortality very greatly, but if the constitutions of the patients had been the cause of the great difference, it could not have reduced them to so great an extent. He assumed, therefore, that there was a difference in favour of the cottage hospitals. What was the cause of that difference? The statement by Mr. Callender, that in former days the mortality after amputation in country hospitals was 17·5 per cent., was very striking. It was very extraordinary that in the old days the mortality should have been less than it had been recently. He had obtained some statistics which brought the fact out very strongly. In the Peninsular War, after the battle of Vittoria, large hospitals were formed, to which the wounded were sent. Hospital gangrene broke out, and caused an immense mortality. Afterwards the practice was adopted of treating the wounded chiefly with the regiments in the field. A certain number were sent away, but others remained with the regiments, and the amputations took place in the field, and the wounded remained there until they were cured. As was the custom in treatment then, the men in the field had spare allowance of food, and few comforts, but the results were these, that of 15 cases of amputation at the shoulder joint 1 died, or 6·7 per cent.; of 98 cases of amputation of the upper extremity 5 died, or 5·1 per cent.; of 84 of the lower extremity 19 died, or 22·6 per cent. These were cases of soldiers absolutely treated in the field, and were to some extent under similar conditions to men in cottage hospitals. After the battle of Toulouse the wounded were placed in hospitals in that town. Among them there were 41 cases of primary amputation of thigh and leg, and 7 of arm, and only 10 died altogether. He would assume that the whole of the 10 were cases of thigh and leg amputation. After the battles of Quatre-Bras and Waterloo there were a number of cases of primary amputation brought from the field to the hospitals in Brussels, others were performed in the hospitals themselves, and of these there were altogether 43 primary amputations of the arm and forearm, of which 5 died, or 11·6 per cent.; 97 cases of thigh and leg amputation, of which 26 died, or

26·8 per cent. Continuing then these groups, it is found that of 148 cases of amputation of the arm and forearm, 13 died, or 6·8 per cent., of 222 cases of amputation of the thigh and leg 55 died, or 24·8 per cent. Altogether there were 370 cases, of which 65 died, or 17·6 per cent., which was just about the rate of mortality that Mr. Callender mentioned as having prevailed formerly, though he did not specify the proportion in which the different cases occurred. In the Crimea, from the 1st April, 1855, when the soldiers were well housed, and had plenty of food, with a liberal supply of stimulants, there were 148 cases of primary amputation of the arm and fore-arm; of these 23 died, or 15·5 per cent., which was rather more than double what took place in the Peninsular war under much worse circumstances. There were 235 cares of primary amputation of the thigh and leg, 118 died, or 50·2 per cent. Thus twice as many died in 1855 as in the latter part of 1813-15. This result took place notwithstanding the use of chloroform, and what were regarded as general improvements in the treatment. The question therefore arose whether all the so-called improvements were really advantageous. The reply must be that some of them were posi-tively injurious.

Mr. H. MONCREIFF PAUL failed to see the connection which the remarks regarding the medical profession made at the conclusion of Mr. Burdett's paper had with its main subject. Mr. Burdett had no doubt good reasons for taking up the cudgels on behalf of that profession, but he (Mr. Paul) was at a loss to follow the logic of the writer in the reasons which he had adduced in support of the proposition, that because the highest honours had been conferred on those who destroyed, and not on those who saved life, it was to the honour of Mr. Lister and Mr. Spencer Wells, but to the dis-honour of the nation, that neither of those gentlemen had received any public recognition from those who distribute the national honours and awards. He must also join issue with Mr. Burdett in the assumption that the medical profession was shunned because its members had not before them the incentive of ultimately securing a seat in the House of Lords. In the choice of a profession men naturally looked to their own fitness and to the monetary advantages which it was likely to carry with it, and the barrier to many in adopting the medical profession was not the absence of honours, but of the *mens sana in corpore sano*.

Professor LEONE LEVI pointed out that the paper gave no dates, and as medical systems changed from time to time, it was necessary to have the date in which any particular case occurred. The pre-ference given to the cottage hospitals as against the larger hospitals would seem to be opposed to all economic principles of working on a large scale as compared with working on a small scale. The larger hospitals had greater resources at command, much greater facilities for the expenditure of capital, in the general treatment of cases, and also a greater choice of eminent surgeons. In addition to this the smaller hospitals had this disadvantage : that they had less public supervision, and one could imagine that where there was not supervision there would be a greater inducement to negligence.

Mr. T. MOORE, F.R.C.S., said that there could be no doubt that some time ago, before Mr. Lister introduced his antiseptic treatment, the mortality of some hospitals was very great indeed. The mortality in the Glasgow hospital, in which Mr. Lister himself practised, was so great that the building had to be pulled down, and a new one erected. The reason was no doubt that the air in a large hospital ward was more or less full of germs of disease emanating from the patients. In small hospitals or private practice it was impossible to have these germs to any great extent. When he wrote the letter quoted in the paper, he felt sure that a small hospital was superior in that respect to a large one, but since then he had come to a slightly different conclusion, namely that it was not so much the size of the hospital that made the difference in the mortality as the size of the ward. In the Petersfield hospital the largest ward contained only two beds, and such a thing as erysipelas or pyæmia was unknown there, except in one instance, in which it could be traced to a surgeon who had been attending other patients in other places. Small wards were also advisable because of the effect produced upon the patients' minds by other patients dying in beds close to them.

The Rev. ISAAC DOXSEY said that to him it was an insoluble mystery why with better scientific training, with greater mechanical advantages and appliances, with every attention to structural and sanitary arrangements, there should be a greater mortality among patients in large hospitals now than in former times. He had endeavoured to account for it by this idea, that perhaps the capacity of the human system for enduring disease was not so great as it formerly was. He was inclined to think that there was in operation some cause deteriorating the tenacity of life. He drew attention to facts stated in Dr. Steel's "Howard" essay, with regard to St. Thomas's Hospital. It was there stated that when the patients from the old hospital were transferred to Surrey Gardens the mortality increased, and in the present hospital on the Embankment the mortality was still greater, notwithstanding that the hospital was supposed to have been built with every modern appliance. It was remarkable that the increased death-rate was more in the male than in the female sex, and more in medical than in surgical cases.

Mr. F. WRIGHT thought it was of vital importance that the hospitals should be kept clean, and he therefore asked Mr. Burdett if he could state whether at the cottage hospitals more attention was paid to cleanliness, ventilation, repainting and rewashing of the walls, &c., than in large hospitals.

Mr. A. E. CLARKE asked whether the antiseptic treatment had been introduced into London hospitals to the same extent as in the German hospitals.

D 2

Mr. BURDETT, in reply, said that statisticians, like other people, could deal only with the figures placed in their hands. It was easy enough to manufacture statistics, but that was an occupation which no scientific man could follow. It was quite true that in the tables given, the number of cases did not in each instance exactly tally, but Mr. Lawson had shown that after rearranging the various classes of operations in exact proportions, there was still a difference of 7·9 per cent. in favour of cottage hospitals. In the tables which gave a summary of all the cases, the difference was 7·1 per cent. in favour of cottage hospitals, so that if Professor Erichsen's cases and his own tallied numerically, in every group of cases the difference would only be 0·8 per cent. Hence the two methods of classification were practically the same as far as results were concerned. The fact that in old days the mortality in large hospitals was less than at present, might be explained in a measure by the knowledge that in former times the arrangements with reference to drainage and so forth were not of a highly scientific character. The drains for instance were not connected directly with the wards, and the excreta and other matters were carried outside the building at once. In the case of field hospitals it was invariably necessary that this should be done, and there was no direct connection with any closed retort, i.e., a cesspool or level sewer, which would generate foul gases. Further, it had been shown over and over again, that for surgical purposes tent hospitals were the most desirable, because there was a rapid renewal of air. Nor were tents necessarily cold, for in Russia, by a system of heating from below, it had been found possible to maintain any required temperature. When an epidemic of blood poisoning prevailed in the Royal Infirmary at Manchester, an examination of the condition of the drainage exposed such a state of horrors that it was found necessary to stop operations altogether for a time, and it was decided to erect ordinary field hospital pavilions in the grounds adjoining. The results obtained there had been satisfactory as compared with those attained in the wards of the infirmary itself. One speaker had thrown doubt upon the logic of the concluding remarks in the paper. The subject for consideration was the relation of the acts of the medical profession to hospital mortality, and when in the course of the paper it was incidentally shown what the labours of two members of the profession had accomplished, he did not consider it out of place to draw attention to the well known fact that medical men occupied an exceptional place in the absence of state honours. Professor Levi had asked for dates. The first cottage hospital was established at Cranleigh, Surrey, in 1859, and the operations recorded in the paper were all of which a record had been kept in the cottage hospital case books during the twenty years ending 1878. He would take care that the dates were especially noted before the paper appeared in the *Journal*. The relative cost per bed in cottage hospitals was considerably less than the cost in large hospitals in towns. Owing to the personal interest that was taken in these institutions by the village residents, he believed that there was far greater public supervision in cottage hospitals than in large hospitals. As an instance of the good results arising from careful supervision, he would mention the Birmingham General Hospital, which was managed by laymen who exercised continual watchfulness. The consequence was that the economy there was a wonder to some of those who had been connected with other large hospitals. Mr. Moore's contention, that the smallness of the wards was of greater importance than almost anything else, seemed at first sight to have something in it, but unfortunately experience proved the contrary to be the case. One of the hospitals in London with nearly 300 beds, had 80 small wards. There were 148 surgical beds, but in 1877 operations had to be discontinued there altogether for a time in consequence of the high mortality. At the Rotunda Hospital, Dublin, there were comparatively small wards. Only one at a time was occupied by lying-in women, the one last occupied being kept vacant for a time before new cases were brought into it. At first the mortality decreased, but it was doubtful whether the system was as efficient as it was expected to be. At the Exeter Hospital the mortality after amputations was small. This result was attributed by the medical staff to the fact that operation cases were always isolated in a ward shut off from the other parts of the hospital. The patients were retained there for at least a week before they were allowed to go into the larger wards. With reference to the depressing effect on other patients of one patient dying in a large ward, he knew of an instance in which it was decided to have a dying ward to which patients were to be removed when all hope was lost of their recovery, but when the first case occurred the doctor refused to take the responsibility of ordering the patient's removal. With regard to Mr. Doxsey's observations about St. Thomas's Hospital, he had no doubt that the large mortality there and at the Leeds Infirmary arose from the faulty construction of the buildings which were so arranged that there were large *culs-de-sac* in which foul air accumulated, and through which a current of fresh air seldom or never passed. He was not prepared to say the wards of cottage hospitals were more frequently renovated than those of large hospitals, but the personal interest that was taken in them caused a constant polishing up to be going on, and the walls instead of being painted were usually papered and varnished. At the Home Hospital for paying patients, Fitzroy House, (16) Fitzroy Square, the walls were papered and varnished. After operation cases the wards were immediately cleansed and left vacant for forty-eight hours if possible, always for at least twenty-four hours, and there had never been the slightest sign of any blood poisoning. Leaving the wards unoccupied and ensuring the free circulation of air were important factors in getting rid of infection. Mr. Clarke said that on the whole the figures given in the paper were disparaging to large hospitals, but they were a fair and true statement of the facts. No disparagement was intended, quite the contrary, but the figures must carry their own interpretation with them.

The Administration of Hospitals and Charities.

WORKS BY HENRY C. BURDETT.

Fellow of the Statistical Society, Honorary Secretary Home Hospitals Association for Paying Patients, late Secretary and General Superintendent of the Queen's Hospital, Birmingham, and the "Dreadnought" Seaman's Hospital, Greenwich.

HOSPITALS AND THE STATE: Hospital Management,
and Hospital Nursing. With Statistics of the Chief Medical Institutions in Great Britain and Ireland.

BOSTON MEDICAL AND SURGICAL JOURNAL.—"Mr. Burdett is one of the highest authorities of Great Britain on the subject of Hospital Administration."

LANCET.—"We are strongly of opinion that there is no social question which presses more urgently for solution than that of the right direction of our hospitals and infirmaries. We regard the paper of Mr. Burdett as a welcome contribution towards the understanding of the question in the popular sense. The tables, grouping together hospitals in classes, show the entire cost of their management and maintenance under certain specified heads; they cover a good deal of ground, and have evidently been constructed with conscientious care and industry."

BRITISH MEDICAL JOURNAL.—"The only accurate statistics of hospital income and expenditure, prepared upon an identical basis, which have ever been published."

PAY HOSPITALS AND PAYING WARDS THROUGH-
OUT THE WORLD. Facts in support of a Re-arrangement of the English System of Medical Relief.

BRITISH MEDICAL JOURNAL.—"We commend this book to all who are interested in the improvement of medical relief—and which of us is not? It is clearly and pleasantly written, it is full of facts, and it contains many valuable suggestions. This work cannot fail to stimulate the reforming movements which have been gathering strength in this country during the last few years."

THE AMERICAN PRACTITIONER.—"Mr. Burdett displays and discusses the whole scheme of hospital accommodation with a comprehensive understanding of its nature and extent, and he does it in fulness without prolixity, and in a clear catholic spirit with perspicacity. The book is a stepping stone, a valuable contribution in the way of introduction to a review and candid reconsideration of the whole subject of legal and organised charity—a theme which much demands reconsideration and re-adjustment in the whole civilised world. A good and timely book and suggestive."

COTTAGE HOSPITALS: General, Fever, and Convales-
cent: their Progress, Management, and Work; with Chapters on Mortuaries and Convalescent Cottages. The Book contains the ground plan and elevation of the best constructed British Hospitals and Medical Institutions having fifty beds and under. Second Edition now ready.

LANCET.—"All who are interested in hospitals owe Mr. Burdett a debt of gratitude for this valuable statement of facts."

SANITARY RECORD.—"We do not hesitate to say that, as a text-book of practical reference, it ought to be in the hands of everyone interested in the subject of hospital management, and it is gratifying to know that this opinion is shared by a very large section of the medical profession."

SPECTATOR.—"At a time when so much discussion is going on as to the relative merits of large and small hospitals, it cannot but be satisfactory to notice the appearance of a good and reliable work on cottage hospitals, by one whose experience in such matters entitles him to speak with more than ordinary authority. Every subject has been treated of, from the building of the hospital itself to the diet of the patients."

THE RELATIVE MORTALITY AFTER AMPUTATIONS
IN LARGE AND SMALL HOSPITALS, AND THE INFLUENCE OF THE ANTISEPTIC (LISTERIAN) SYSTEM THEREON. With many Statistical Tables.

LANCET.—"However sceptical we may be as to the sufficiency of any possible statistics for deciding the true relative mortality, after amputations in large and small hospitals, it is easy to recognise the value of the facts bearing upon this subject contained in this paper. They must have been collected with infinite labour, and represent much additional information, for which all interested in the subject owe a debt of gratitude to Mr. Henry C. Burdett, the author."

J. & A. CHURCHILL, 11, New Burlington Street, London, W.

www.ingramcontent.com/pod-product-compliance
Lightning Source LLC
Chambersburg PA
CBHW022031190326
41519CB00010B/1672